# The Adventures of Jackson's Cucumbers

Book #1 By Amanda Green

This literature is a work of fiction. All names, characters, places, and incidents are products of the author's imagination.

All rights are reserved, and no part of this book may be reproduced in any form without the permission in writing from the author, except by a critic who may quote brief passages in a review.

Copyright 2022 GreenBookEats Publishing ©

## Acknowledgments

*"My muse inspires me and pushes me to be great. My muse loves me and direct me to go straight. I may even slow down, but my muse says, "don't quit," I may not write for months, I'm not ashamed to admit. But with his unconditional love and constant encouragement, my muse forces me to keep going and sometimes even sprint. To my loving husband whom I call Green, I thank you for always believing in me and being my muse machine. I love you dearly."*

*"This book is dedicated to my four beautiful daughters. Take the necessary steps to achieve your goal and eventually it will come to fruition. I love you girls."*

Have you ever felt like something great was going to happen? That's exactly how Jackson felt as he rolled out of bed early on a Saturday morning. It was the first day of summer and Jackson had a summer goal.

On the last day of school, Jackson's teacher Ms. Mack asked, "what are your plans for the summer Jackson?" Jackson replied, "I am going to be a farmer and grow cucumbers!" Jackson was so excited about growing cucumbers.

Jackson thought about the day before how some kids laughed at him because they had never heard of an eight-year-old farmer. However, Jackson was determined, and his mom was also determined to help him accomplish his goal of becoming a farmer and growing cucumbers.

Right after Jackson ate lunch that day, his mom drove him to the local Depot to buy cucumber seeds, potting soil, flowerpots and a small white fence. Jackson learned from the internet what he needed to start a garden, so he knew exactly what to purchase.

As soon as Jackson got home, he put on his farmer outfit and quickly went to work. Jackson dumped the cucumber seeds into several flowerpots and covered the seeds with potting soil. It had to be about thirty seeds total. Jackson was anxious to see how many cucumbers he could grow.

Jackson watered his cucumber seeds everyday. He learned from the cucumber package that it would take fifty to seventy days to grow cucumbers. Jackson had a plan. Once the cucumber seeds sprouted into a plant, he was going to remove the plants from the flowerpots and bury them in the ground for the cucumbers to continue to grow from his garden.

After about four weeks of watering his cucumber seeds, Jackson finally saw small leaves in his flowerpots. Jackson was so excited and yelled for his mom to come see his small plants. Jackson's mom quickly came outside to see what the excitement was about. It was now time for Jackson to uproot his plants from the flowerpots and bury them in the ground to start his garden.

Jackson's mom helped him find the perfect location in the yard and watched how happy Jackson was as he watered his cucumber plants. Jackson was smiling from ear to ear because he was hopeful that he would have cucumbers very soon.

After about two months, Jackson discovered he had six yellow flowers each housing tiny cucumber bulbs. Jackson was so anxious for his cucumbers to grow bigger! He even moved his white fence further away from his garden to give it more room to grow. Jackson continued to water his garden daily.

One morning Jackson woke up bright and early and ran outside to check on his cucumbers. Jackson was so eager he was still wearing his stripe green pajamas. A few days had passed since he watered his cucumbers, but Jackson was still happy and hopeful that his cucumbers would be bigger by now.

To Jackson's surprise he only had four out of six cucumbers remaining. Just in a few days, two tiny cucumbers had fallen off the plant and died. Jackson could not understand how his cucumbers were dying and not growing. Jackson quickly ran inside to tell his mom the bad news.

Jackson kept watering his cucumbers but after one week, he discovered another cucumber had died. Jackson now had only three tiny cucumber bulbs remaining. Jackson was now worried and knew he had to figure something out quickly before all of his cucumbers died. His mom's voice echoed in his head.

The next morning, Jackson began to research ways to save his last three tiny cucumbers. He started with Google and typed in the search box "how to make cucumbers grow." Amongst the information that Jackson found, he learned that cucumbers not only needed a lot of water but required pollination. Jackson also learned that in order for pollination to take place he needed bees.

Jackson ran to his mom with excitement, "mom, mom, I figured it out! We need bees!" Jackson's mom said, "calm down Jackson, why do we need bees?" However, Jackson could not calm down, because he had just learned how to save his last three cucumbers. "According to Google, the bees will transfer pollen from the female flower to the male flower for the cucumbers to grow bigger."

Jackson was so distracted by the colorful flowers he walked right past the garden employee while talking. Jackson was talking fifty miles per hour, *"my---cucumbers—died—no—bees—and"* it sounded like a bunch of gibberish but oddly, the garden employee understood exactly what Jackson was trying to say and he knew exactly what Jackson needed.

He handed Jackson the most beautiful golden potted flower with a sticker that read *"evening primrose."* "Here you go young man, these will surely bring the bees and even butterflies." Jackson quickly purchased six potted beautiful yellow flowers and returned to his car where his mom was waiting.

After Jackson made it home, he quickly changed into one of his favorite farmer outfits and planted the evening primrose flowers right next to his cucumber plants. Jackson made sure the flowers had plenty of water.

After about two weeks, Jackson woke up bright and early and felt like something great was about to happen. Jackson ran outside while still wearing his blue pajamas and discovered he had three healthy growing cucumbers. He even poked at the friendly bees and butterflies dancing across his entire garden.

Jackson quickly got dressed and began taking pictures of his garden to share on his Instagram page. Jackson's mom monitored Jackson's Instagram page for safety reasons since he was so young. In a matter of one hour, Jackson had more than fifty likes and over twenty positive comments on his Instagram post. His favorite comment of them all was left by his teacher Ms. Mack, "great job Farmer Jackson, I love your cucumbers!" Ms. Mack's comment made Jackson's day.

Farmer Jackson continued to nurture his beautiful garden, which now had four healthy growing cucumbers. Jackson was full of joy because he had finally accomplished his goal of becoming a farmer.

# Comprehension Questions

Answer the questions below using a separate sheet of paper.

1. What is the main idea of the story?
2. Who are the characters?
3. What is the setting of the story?
4. When does the story take place? (hint: what season?)
5. Why was Jackson's cucumbers dying?
6. Was Jackson's mom supportive, explain how do you know?
7. How did Jackson feel when he discovered his cucumbers died?
8. How did Jackson find the solution to grow his cucumbers?
9. Based on the context clues, what does pollination mean?
10. Based on the context clues, what should you do if you need to learn information about a subject matter?
11. How did Jackson learn what he needed to create a garden?

*AmandaTheAuthor* is originally from Fort Pierce, Florida. Today *AmandaTheAuthor* lives in Florida with her husband and four beautiful daughters. She works for one of the largest Clinical Research Organizations in the United States, managing Clinical Research Associates. In her free time, she enjoys volunteering at her church, taking photos of others, and simply writing. Check out other books written and created by *AmandaTheAuthor*: "Express Yourself, Poetry Activity Book; Just for You Mom, Word Search; Just for You Auntie, Word Search and Just for You Grandma, Word Search." Keep an eye out for more books about "The Adventures of Jackson." Feel free to contact *AmandaTheAuthor* at info@greenbookeats.com or on social media @AmandaTheAuthor for answers to the comprehension questions and to request Jackson's unique word search sheets.